GOD OR NOT?

Faith,
Hope you like
this book. Goes
back to days of
my friendship
to Art.

David H.
12/31/21

GOD OR NOT?

One person's amazing experiences:
Do they verify the existence of God…or not?

DAVID S. HEEREN

God or Not?

www.endtimesmystery.com

1603 Capitol Ave., Suite 310 Cheyenne, Wyoming USA 82001
1-888-980-6523 | admin@urlinkpublishing.com

URLink Print and Media is committed to excellence in the publishing industry.

Published in the United States of America

Library of Congress Control Number: 2021904236
ISBN 978-1-64753-718-0 (Paperback)
ISBN 978-1-64753-719-7 (Digital)

11.02.21

PREFACE

This small book is about one of the most important things in my life and yours: Religion. There are many religions in the world, but ultimately they condense to three central issues:

- Does God exist?
- Can I have a personal relationship with him?
- Where will I spend eternity?

My outlook on life and yours depend on our choice of answers to these questions. Those who believe there is a personal God, in whose presence we shall live forever, have a hopeful outlook.

Those who do not – atheists, deists, earth worshipers, New Agers, et al – if they face the issue logically, are hopeless human beings. No personal God implies no afterlife. No afterlife means the flawed present existence is all.

I am an optimist. I believe I can and do have a personal relationship with the God who has created and sustained planet Earth, its solar system, and the galaxies and universe beyond.

I believe this not because I am in a condition that a psychiatrist might describe as denial. I believe this because the evidence of many events of my life bears out its truth. These are recounted tersely and accurately in this volume.

Twelve foundation blocks in the construction of my faith are recounted herein. If the reader lacks such a faith, believing that God does not exist and that he or she is a product of evolutionary random chance, this book can be of value as an avenue to new truth.

It is certainly an avenue to faith, hope and the highest character trait to which a human being may aspire: Love.

I heard about love in high school and college. But, with no exceptions that I can remember, my teachers were concerned more with its carnal elements than the spiritual ones: Their philosophy was self-centered, not God-centered.

Perusal of this little book could broaden the horizons of hope and love for open-minded readers. Everything it describes was either inspired by God or coincidental. The possibility of coincidence diminishes as the evidence mounts.

David S. Heeren

CHAPTER ONE

'GO AHEAD AND QUIT'

> "Bring the whole tithe into the storehouse, that there may be food in my house. Test me in this," says the Lord Almighty, "and see if I will not throw open the flood gates of heaven and pour out so much blessing that there will not be room enough to store it."
>
> (Mal. 3:10)

I am trained in Evangelism Explosion. An important part of the EE presentation to an unbeliever is a personal testimony. The Christian is supposed to explain to the unbeliever what life was like before and after he/she became a believer.

I had a problem with this because I do not know when I began to believe. This means my salvation testimony lacks the effectiveness it should have. However, my life has been full of God's faithfulness, and the fiscal testimony recounted in this chapter is amazing even to me.

It began when, at the age of nine, I began collecting baseball cards sold in packages of chewing gum.

I was a huge baseball fan. After school I often played my own fantasy games and announced them so enthusiastically that neighbors would complain to my parents that I had a radio turned on too loud.

When they learned I was a "major league baseball announcer" they joked about it good-naturedly with my parents. But I was serious about it and some 20 years later, I did get a chance to work a dream job as public address announcer for the New York Yankees during spring training in Fort Lauderdale.

For me, collecting baseball cards was a hobby I treasured because of my childish fanaticism for the Yankees overlapping the championship Joe DiMaggio-Mickey Mantle years.

I collected the cards for five or six years and stored hundreds of them, in perfect condition, inside shoeboxes. I had several boxes full of the treasured cards, including at least one Topps Mantle rookie card. This card, in mint condition, is selling today at auction for more than $150,000.

I remember in one package of five cards I bought there were three Ted Williams cards. A single non-rookie Williams did not equal in value a Mantle rookie card, but the total value today of the three Williamses equals or exceeds that of a single Mantle card.

Full value of my collection on today's market would be worth several million dollars. All I had to do was keep the cards and I was guaranteed to become a wealthy man by exercising just a modicum of fiscal common sense. I was so fanatical about collecting the cards as a youngster that there was no way I would willingly give them up at that time.

But there came a crisis point when I graduated from Teaneck (NJ) High School in 1956 and enrolled at the University of Delaware. A short time after my enrollment, my grandmother Seibert became terminally ill and my widowed mother sold the house in Teaneck and began packing for the move to Catasauqua, PA., where she would take care of my grandmother until her death.

Before initiating the moving process, my mom telephoned me and asked if I wanted to keep anything. She hinted that my baseball card collection was something she wanted to throw away. At that time, nobody knew how valuable those cards would become.

I wanted to keep the cards and should have told my mother so. Instead, deferring to her rigid tone of voice, I said compromisingly: "I'll let you know."

But the next time we talked, she told me she had given the cards to a neighbor boy. I was disappointed, and it was an emotion that intensified through the years as the cards became more and more valuable.

About twenty years later, during the awful economy of the late 1970s, my Christian faith hit bottom. My wife Joan and I had two sons in elementary school when Joan told me one day that she wanted to quit her job so she could spend full time at home with the boys.

She was a registered nurse in charge of a hospital floor, so her income was about equal to mine. I knew without doing a single computation that our budget could not stand a drop in income of fifty percent. After discussing the matter, we agreed to seek counsel from the pastor of the church we were attending.

After listening to us place our conflicting interests before him for about half an hour, the pastor, in his usual combination of blunt candor and faith, said to Joan, "Why don't you go ahead and quit. The Lord will provide."

The Lord did provide, but it didn't happen nearly fast enough to satisfy me. Within a year we fell behind in our mortgage payments. A bank turned down my request for a loan.

I'm not proud of the next incident, but it is part of the story. I thought the bank had blown me off without seriously considering my loan request. Since I had consistently paid my bills on time for twenty years, I thought I deserved better. I told the Lord so, and mentioned my displeasure with the bank in that prayer.

A few weeks later, while driving past the bank's main office, I noticed a sign posted stating that it was permanently closed. I've always wondered if that was mere coincidence or if my prayer had been a factor.

But the bank's status was not as big an issue to me as the looming homelessness of my family. Neither my mother nor Joan's parents were well-to-do, so my pleadings to the Lord became louder and longer.

Finally, on a Wednesday, I received the dreaded letter from the bank that held our mortgage and was still in business: Make at least one mortgage payment by the next Monday or foreclosure proceedings would be initiated.

I grappled mentally with the problem, but it burrowed in like a mole. *You made a big mistake, and there is nothing you can do about it now.*

I spent most of my time worrying. I knew the Bible advised "Be anxious about nothing." (Php. 4:6) But no matter how hard I tried to climb the hill of faith, I kept sliding back into the mire of apprehension.

The situation seemed hopeless. I could not come up with the money or even a clue of how to get it. The rejected bank loan seemed to have been the last hope.

Friday was payday, but most of that money had to be committed to necessities such as food. The rest did not constitute nearly enough for a mortgage payment. The full paycheck was not large enough to cover a single payment.

I told no one about the financial plight. Not the pastor, nor the associate pastor, nor anyone else at the church,

Meanwhile, I struggled with a new problem. As a Bible-believing Christian, I felt it was my duty to God to contribute a tithe of all income to the church. But suppose I skipped a week? I could make up for it some other time. If I kept the tithe money in my wallet, I'd be that much closer to being able to make the mortgage payment.

But I'd still be short.

I turned in my Bible to the book of Malachi and read: "Bring the whole tithe into the storehouse, that there may be food in my house. Test me in this," says the Lord Almighty, "and see if I will not throw open the floodgates of heaven and pour out so much blessing that you will not have room enough for it." (Mal. 3:10)

My decision was made. On Sunday morning, I placed my tithe check in the offering plate.

Then I went right back to worrying. I worried through the rest of the morning. I worried all afternoon. In the evening, Joan and I returned for another church service and I managed to worry all the way through it.

After that service, the associate pastor walked up to me, smiled and handed me an envelope. "The Lord told me to give this to you," he said, then turned and walked away.

In the envelope was a check for the exact sum needed for the mortgage payment.

I don't know how the associate pastor became informed of my problem or the sum to the exact cent that I needed. What tone of voice did God use when he "told" him to give me the check? I don't know why God did anything about it at all. It was not his problem.

Well, maybe it was: I couldn't read his mind.

But I did know this: I tithed and God kept his promise by pouring out a blessing.

From that Sunday more than forty years ago to the present time, no one in my immediate family has encountered serious financial difficulties. A surprise inheritance check from my mother's estate enabled us to buy a home outright. We have not made a mortgage payment of any kind since 1995.

My faith and Joan's were greatly strengthened by these happenings.

But even these things were just snapshots in comparison with the big picture. I have shared this testimony of God's faithfulness numerous times. The response is always the same: Amazement at the provision of God and his love.

It was printed in a *Chicken Soup for the Soul* book entitled *A Book of Miracles*. The reaction was outstanding. For instance, a man living in the Middle East contacted me via e-mail. His commendation was so sincere that I thought he must be a Christian.

After several exchanges of messages, he informed me that he was a Muslim, but his positive attitude reminded me that not all Middle-Eastern Muslims are enemies of American Christians.

In fact, I have learned from missionary sources that the Gospel of salvation through faith in Jesus Christ is being accepted by thousands of Muslims living under oppressive regimes, especially, ironically, in Iran and Iraq.

My fiscal testimony has been told and retold many times through various media and is being retold through this book.

Within the last two months, two Christian friends have asked me to pray about their financial problems. One lost a job; the other lost a lucrative contract, leaving both uncertain of their fiscal future.

I shared this testimony with both, and they thanked me saying they were greatly encouraged by what I told them.

But what about those baseball cards? Suppose I were a multi-millionaire thanks to income from selling them: What then? Well, for one thing, this testimony would not exist and could not have been read by strangers halfway around the world.

I have no idea how many people have been encouraged by it, or how many have been strengthened in their faith, as have I. To me faith in Christ is now unconditional.

Vast wealth doesn't seem important to me any more compared with other blessings God has poured out.

CHAPTER TWO

DO YOU BELIEVE?

The sudden pain struck with dagger force, twisting deep into my lower abdomen. I doubled over and had to steer the car to the side of the road. Joan exchanged seats with me and drove to the nearest hospital.

In the emergency room, between spasms of pain, I explained my symptoms to Joan. She was a registered nurse, so I asked her for a professional opinion. She told me it was likely that I had colitis.

I wondered: *What if it's colon cancer?*

It was late in the day when I finally settled into a hospital bed. I lay there for many hours without being able to sleep. I was still experiencing pain, even though I had been given a painkiller. But the thing that kept me awake was anxiety.

I believe in the power of prayer, but not until after midnight did it enter my head that I hadn't talked to God about my illness, whatever it was.

I began talking as if to him, silently, in my conscious mind: "Lord, this is awful. I'm scared. I've never been this sick in my life."

I continued to pray, and then, suddenly, a thought occurred to me: *Do you believe I can heal you?*

The thought was in the first-person tense, as if someone else had established a presence in my mind, but it seemed to me that it was a thought of mine.

I answered the question affirmatively in my mind. And, immediately, another thought took its place: *Do you believe I want to heal you?*

I smiled over the fact that I seemed to be having a mental conversation with myself, but the smile was brief. This was a conversation, whether it was one-sided or two-sided, and I needed to take it seriously.

"Yes, Lord," I said internally. "I have read the Bible I know you are a merciful God. You have raised the dead. You have cured many sick people."

Then, one more thought: *Do you believe I will heal you?*

I wrestled with this question for what seemed a long time before stating my answer succinctly: "Yes, Lord, I believe you will heal me."

And then I fell asleep.

When I awoke the next morning, my body was suffused by a feeling of serenity. The pain was gone. Medical technicians put me through a series of tests lasting all morning, but I believed the tests would turn out well.

In the afternoon, when a doctor entered my room, he was frowning. I stiffened, preparing myself for bad news. But I was not frightened. There was within me a confidence that seemed inexplicable even in the presence of the grim physician. The fact that I was pain-free probably had something to do with it.

Spreading his hands and shaking his head in a gesture of helplessness, the doctor said: "I have never seen an overnight change like this and I don't understand how it could have happened, but there is no reason for you to stay here for another day because there is nothing we can do for you. No, correct that: There is nothing we *need* to do for you, because there is nothing wrong with you. I'll place the order for your discharge."

Still shaking his head, he walked away without saying another word.

(This story appeared in a *Chicken Soup for the Soul* book entitled *Devotional Stories for Tough Times*.)

CHAPTER THREE

VISITATIONS

> The Angel of the Lord encamps around those who fear him, and he delivers them.
> (Psalm 34:7)

There have been two events in my life involving the protective presence of beings that seemed to be angelic. In one of the two I saw the being.

I was eleven or twelve years old when I saw something that looked like an angel during a Boy Scout camp-out in northern New Jersey near my home.

It was late at night when I awoke. I pushed back the flap of my tent to admit some fresh air and noticed a bright light hovering near the front of the encampment.

As I watched, my eyes focused on details of the light, which appeared to be a human-shaped object.

Captivated by the unusual sight, I continued to watch the light for at least a half-hour. From time to time there was motion within the light, as the human-shaped being, whatever it was, changed position. It appeared to be seated on a low tree branch, with its back turned in my direction.

When the thought first occurred to me that it might be an angel, I dismissed the idea. But the longer I watched, the more my thought pattern changed from doubt to near certainty that at least one angel was keeping watch over the encampment.

The angel, if that is what it was, remained in a position facing away from the camp as if to be prepared should anything dangerous approach. It did not occur to me at the time to look in other directions to see if any other "lights" were visible on the perimeter of the camp. But from time to time I did detect clear movement of the illuminated humanlike form I was observing. It did not have wings.

I spent enough time checking out the other tents in the encampment to be sure that I was the only one observing what I came to believe was a supernatural phenomenon.

Was it an angel? Was God preparing me for other experiences of similar origins which He knew through divine foreknowledge that I would encounter during my time on earth?

When my attention wandered from the light and I became drowsy, I closed the tent flap and returned to my sleeping bag. The thought occurred to me that maybe my parents and/or parents of some of the other scouts had prayed for us and God responded by sending a protective angel to our camp.

It was not a frightening experience, but one that gave me a peaceful feeling. I fell asleep quickly after crawling back into my bag.

I think I saw an angel that night, but do not know for sure. I might have been mistaken. Of the second experience, twenty-five years later, I have less doubt, even though there was no divine manifestation to be seen.

At that time Joan and I had two sons ages six and nine. The older boy, Michael, was named for an angel identified in the Bible, and on that day I believe a being akin to Michael's heavenly namesake performed a supernatural feat to spare our family from tragedy.

We were living in the Fort Lauderdale area and often spent Saturdays picnicking at a beach we had found that never seemed to be crowded. After packing a picnic lunch, we joined hands in prayer as we did routinely before making a drive in heavy traffic. Upon later reflection, I have reason to believe that a divine response to our prayer may have saved the lives of Michael and Daniel.

From where we lived in Plantation a few miles west of Fort Lauderdale, the drive to our favorite beach on a peninsula in Dania required us to spend about ten miles on I-95, which seemed then and remains to this day a veritable race track for reckless drivers.

It was during the recession of the late 1970's, shortly after Joan became a stay-at-home mom. We had a single-salary family income and drove used cars.

The automobile of choice that day was an American Motors Gremlin.

A generation later, in two separate polls, the Gremlin was labeled "one of the fifty ugliest cars of the past fifty years" and "one of the twelve worst cars ever built." In a collision with another motor vehicle it had the potential to crumple like an accordion.

The Gremlin was a subcompact with an almost vertical hatchback. There was no trunk to cushion a collision involving the rear of the car. Such a crash was potentially fatal for anyone sitting in the back seat. That day, Michael and Daniel were in the back seat of the accordion car.

I was a careful driver. Before and since that day, I haven't been involved in an accident with another vehicle in which both were in motion. But that day it didn't matter how careful I was.

On the way home from the beach we were heading north on I-95 in the extreme right lane, driving between 60 and 65 miles per hour. The Gremlin probably couldn't have gone much faster if I had floored the gas pedal, but I was driving conservatively. It seemed to me as if at least 95 percent of the other cars on the road, and even most of the trucks, went roaring past us.

Then I saw it.

In the rear view mirror I noticed a car driving at a very high rate of speed, even for I-95. It was weaving in and out of traffic, zooming past vehicles that were going at least 70 MPH. As the speeding car approached from the rear, it was in the lane to our left. But it closed so rapidly on another car that suddenly, to avoid crashing into the rear of that car, it veered to the right.

It was closing on us so fast that the Gremlin might as well have been parked. I don't know how the driver failed to see us, but the speed of his approach made it obvious that he didn't. Normally, under such a circumstance, I was prepared to drive off the road to the safety of a shoulder. But it was too late.

At the last possible instant, just before crashing into us, the approaching car made a strange move that I can visualize in slow motion to this day. There was no sound of brakes being applied or of squealing tires. But, as if controlled by a powerful unseen hand, the speeding car slowed in an instant to about the same speed as ours and then swerved suddenly to the left. It missed by inches both the rear left bumper of our car and the front right bumper of a car it had just passed.

The wayward vehicle skidded across two more lanes of traffic, narrowly missing several other cars. Then it struck the road's concrete center divider and careened back across the four northbound lanes, just missing collisions with numerous vehicles. It slid off the right side of the road and down an embankment where it came safely to a stop on a grassy strip. It was the only vehicle with a scratch or dent resulting from the incident.

I don't know if young Michael and Daniel realized how close they came to serious injury or worse, but

Joan and I did. As a family, when we arrived home, we held hands and thanked God for sending protection. I remain convinced of the miraculous intervention on that day of at least one unseen being. For me it has been an enduring lesson in the love of God.

CHAPTER FOUR

PRIDE

This book is mostly chronological. This chapter, based on the divine out-workings of pride – especially mine – can be fitted in anywhere because, despite numerous painful lessons, it's still a problem for me.

I'm competitive by nature, especially in athletics. I've always struggled with pride over the outcome of sporting events I participated in or spectated. I was devoted to the New York Yankees as a boy and young man. If they should happen to lose a pennant or World Series, I'd be depressed for weeks.

Mickey Mantle was the greatest athlete – power to hit a baseball more than 600 feet and speed to reach first base batting left-handed in three seconds flat – ever to wear a Major League Baseball uniform. He led the Yankees to 12 American League pennants and seven World Series titles in 14 years, thereby sustaining my youthful ego.

To this day, nearly six decades later, I'm still capable of kicking a fit over the 1960 World Series, in which the Yankees scored more than twice as many runs as their opponents. They lost, even though Mantle was a dominant player.

I have been on the winning side most of the time, competing or rooting for a favorite team or athlete. But all too often the cost has been much greater than the price of admission.

At times the problem has been boastfulness. While maintaining a demeanor of pseudo humility, I've managed to inform everyone around me of personal accomplishments on softball and football fields, running track, golf course, tennis and basketball courts, etc., etc. – you get the idea.

I've bragged about being chosen the outstanding athlete in my college fraternity. I've bragged about placing second in a massive men's table tennis tournament in Fort Knox. I've bragged about leading several men's softball leagues in batting. And about player-coaching a hodge-podge team consisting of men and women employed in a small office to nine victories in ten games, including upsets over two all-male all-star teams.

So there, I've done it again: A whole paragraph full of boasts.

While I was editor-in-chief of *The Review*, the University of Delaware student newspaper, we did not cover fraternity events in our sports section. With our small staff, it was all we could do to cover varsity sports.

But we did cover the fraternity track meet in my senior year. Why? Well, I suppose it was just coincidental that I happened to break the meet record for the mile run by more than 30 seconds that year.

I wrote the story myself and made sure it got into the newspaper.

It was such blatant bad taste that one of the newspaper's staff members angrily confronted me. "That's an exaggeration!" he shouted. And that touched me off. You see, it really wasn't an exaggeration, so I answered back.

While reflecting on the incident later, I came to a sobering conclusion: I was a self-centered human being and a hypocrite. Here I was, a man claiming the faith of Jesus Christ, which decries selfish pride. But my angry response to a detractor proved beyond doubt that I was as self-centered as the most rebellious unbeliever.

There is a point to all of this and here it is: If in fact God exists and is the Author of the Bible, as that book claims, then I should expect reprimand every time I allow a nasty sin such as pride to surface flagrantly in my life. The Bible contains these comments clarifying the Lord's attitude toward pride:

"Whoever has haughty eyes and a proud heart, him will I not endure." (Psalm 101:5)

"To fear the Lord is to hate evil; I hate pride and arrogance, evil behavior and perverse speech." (Prov. 8:13)

"When pride comes, then comes disgrace, but with humility comes wisdom." (Prov. 11:2)

Like the angry response to my ill-advised story about my own track-and-field exploit, there have been consistent recriminations for my most prideful actions and boastings. Here are examples from a few of the many sports in which I participated.

I player-coached a men's softball team in a highly competitive league. Before one game I made an ill-advised remark to the coach of the opposing team who until that time had been my friend. I lost the friend and my team lost the game by something like 20 runs.

My dad was in his forties when I decided to show him how good I was at softball pitching (I wasn't much good at all). I think dad hadn't held a bat in at least twenty years. Anyway, I tossed him one pitch and he hit a shot over a fence into a neighbor's yard that was – take my humiliated word for this – a long distance away.

A few years later, two college fraternity brothers of mine drove with me to Vermont for a weekend of skiing. None of us had skied before then, and I don't know about the others, but I haven't skied since then.

On the first day, after an hour of basic instruction, we got bold. At the end of the day, at twilight, after the other skiers had gone into the lodge at the foot of the slope, we walked halfway up the mountainside and put on our skis.

The most emphasized instruction for beginners is to weave back and forth on ski edges in order to control your speed and avoid disaster. After donning the skis, I got that hotshot feeling, and sped straight down the hill on flat ski bottoms. Now, I know what you're thinking: *I crashed and tore up both knees.*

That's not what happened. God – the only reasonable explanation, in my opinion – had mercy. Though attaining breathtaking speed, I rode all of the bumps to the bottom and swirled to a perfect stop a few feet from the ski lodge.

That night must have been unbearable for my frat buddies. They had to endure one boast after another from me. Since they had tripped over each other's skis at the outset, and neither had finished the run, the thing that seemed to bother them the most was my expressed intention to write about the event, including their embarrassing fall, for the college newspaper.

With vehemence they talked me out of it.

But that's not the end of the story. Next day, I decided not to risk my life again with another high-speed venture straight down the slope. I cruised back and forth at 5 or 10 MPH. To this day, I can't explain how it happened, but I took a scary fall that ended my brief skiing career.

My opinion? God didn't want me dead or crippled, so he saw me through the first wild ride. But he wasn't going to let me get away with being a conceited jerk. Considering the caution I exercised the second day at very slow speed, the fall was a dandy. My body and psyche felt equally damaged.

My frat brothers joined the Almighty with satisfied smiles.

Why do I believe this? Well, the outcome of pride for me is always something like this, and I do mean always. It precedes a fall (Prov. 16:18).

Before Joan and I were married, she invited me to a family picnic. She had a large family and there was a touch football game involving mostly young men related to her. They let me play. My team won by 20 points and I made just about every kind of big play in the book on both offense and defense. After the game, I smilingly asked her if she had seen me do this….or that…or anything.

She shrugged. "I was having a good time, talking to the girls."

After we were married, I invited her to a men's softball game. During the game, I hit the ball as hard as I could. She didn't see that, either.

About 15 years earlier, during a softball game when I was a scrawny kid with not much batting power, a hotshot outfielder came in to play close behind the second baseman. I hit the ball over his head for a triple.

Was there a lesson in this for me?

After I developed a much more powerful swing, my attitude, as I have already confessed, swung over to that of the hotshot. In a game in which a friend of mine I knew did not have much power was playing, I moved in from my centerfield position to about 30 feet behind second base when he came to bat. He hit the ball over my head.

A year or two later, in a coed game, I did the same thing. When a young woman came to bat who didn't look as if she could hit the ball very far, I moved in to play a shallow centerfield.

She hit the ball way over my head.

As a 10-year-old, I must have had pretty good teeth. My dentist told my dad my teeth were practically indestructible. My dad told me. Both of us proudly shared this bit of useless information with other people.

As an 11-year-old, I found out my teeth weren't indestructible after all. While playing a game in which I did some rapid roller skating, I skated face first into a steel pole.

One upper tooth was knocked out. Two others were split in half, as was one lower tooth. The bottom piece had to be surgically removed from its lodging beneath my lower lip. Another lower tooth was permanently slanted inward. I wore three successive bridges to support the wounded teeth for nearly 70 years and now wear a partial denture.

Three dental projects resulting from the original incident cost me thousands of dollars and teeth that no longer looked indestructible: High price to pay for bragging.

My father did not elude divine responsibility either. Emergency medical and dental work right after the injury, followed by creation and insertion of the first bridge, cost him a lot of money, payable during a short period of time.

The one consistent thing about all of these incidents is that the outcome was embarrassing for me. Like the showoff ordered by his teacher to stand in a corner of the classroom, I was humiliated by the outcome every time.

It seems to me that humiliation is the perfect punishment for boastfulness.

All of these incidents might be thrown together onto a pile labeled *COINCIDENTAL*. The reader may feel free to do so, but I don't.

Whatever the verdict, the outcome for me has been very expensive, and I have tangible reminders of the worst incident lingering in my mouth.

CHAPTER FIVE

PRESIDENT FORD

I pray every day for the U.S.A. This nation was not mentioned at all in my petitions to God until I was 37 years old. But it took only three days to convince me to do so for the rest of my life.

How did this happen? Well, on Sept. 20, 1975, while praying for the usual self-centered things, I felt an urge to pray for the protection of President Gerald Ford. I asked God to send angels to protect him.

I didn't know why I said this prayer. Ford was in no peril I was aware of. I did not expect a divine response because at that time I didn't believe that the prayer of a solitary person in a nation of 300-million could have much effect.

The next day I felt I should pray the same prayer again, and the day after that. On that third day, Sept. 22, while listening to an evening newscast, I learned that an attempt had been made in California to assassinate President Ford.

The reporter said it was a miracle that a point-blank shot at the president from about forty feet had missed its mark. A marine witnessing the incident tackled the would-be assassin as she was aiming the gun for a second shot.

One person was injured, but no one was killed. The president was unharmed. It was later reported that the outcome of the assassination attempt had not been miraculous after all because the shooter did not have her usual gun. The one she had was new, and it had faulty aim.

The judge who presided over subsequent legal proceedings said he was convinced the president would have been killed if the assassin had been using the gun with which she was familiar.

As I thought about it and learned more information, I concluded that, whether you called it miraculous or not, God had intervened in answer to prayer that he himself may have inspired.

There were two reasons the assassination attempt had been made with a defective gun, and neither seemed coincidental to me. The shooter had decided before Sept. 20 that she would try to assassinate President Ford.

On Sept. 21 she experienced her first setback when police took from her a 44-caliber pistol with which she was an excellent shot.

Knowing that her only chance to assassinate the president would occur the next day, she hurried out to buy another gun – a 38 caliber pistol unfamiliar to her. She had no time to practice with the new gun.

I don't think it was random chance that the gun she purchased was flawed, causing her shot to miss the president. Very few new guns have aim that bad. I hold a mathematics degree, so it occurred to me to find out how many. I discussed the matter with a dealer who has sold hundreds of guns.

The dealer told me that the chances of a new gun having sights so bad that a well-aimed shot could miss its target by at least one foot from a distance of forty feet is virtually nil. He said new guns occasionally are returned for adjustments, but none he knows of has had sights that bad.

A minimum of one foot is the distance off target a gunshot must be to miss entirely a man of average size whose chest is targeted dead-center.

I'm sure I was not the only person who had felt the urge to pray for President Ford shortly before the assassination attempt. How many were praying I don't know, but I don't think I was alone.

Since that day I have prayed for many concerns at the national level. At least four times the prayers were answered in ways that, to me, persuasively identified God's involvement.

Three were presidential elections. These elections were of such importance and the results so unexpected, that I have come to believe the outcome of all three would have been different had not a significant number of people been praying for the election of God-fearing candidates over unbelievers.

I do know that much prayer was being offered regarding those elections. I know this because I have become associated with three nationwide groups that offer prayers concerning national issues. And there are many other such groups.

One of the unlikely presidential winners was a man who had been predicted to lose by a wide margin in every poll right up to Election Day. In most of the polls the anticipated margin of victory for his opponent was of landslide proportions.

There were surely many people praying over these elections. But I am aware of no evidence indicating that the number praying could have been as large as ten percent of the nation's total population. However many there were, the prayer warriors were decidedly in the minority.

The lessons etched into my heart and mind are that God answers petitions by outnumbered individuals and comparatively small groups; and He cares about the course of nations, including the U.S.A.

CHAPTER SIX

CATS

Even as a young cat Rocky was a mime. You could call him a copy-kitten.

At first, this characteristic showed itself as a doglike trait of following people around the house and mimicking some of the things we did. Rocky would follow at my heels or Joan's or one of our sons, Mike or Dan.

If you walked into a bathroom and turned to close the door to prevent Rocky's entrance, he would stand on hind legs and lean against it from the outside. If the door had not clicked shut behind you, Rocky would shove it open.

The first time Rocky broke into the bathroom while I was standing next to the toilet it was embarrassing. I rushed over and slammed the door, leaving Rocky still inside with me, which was where he wanted to be. He watched my every move in that forbidden room.

The way the bathroom doors in our home were designed, Rocky could not push one open to leave a room, but he enjoyed barging in. He was the only one of more than twenty pet cats Joan and I had through the years that would do this. We had many paws-under-the-door cats, but only one that was repeatedly guilty of breaking and entering.

Rocky broke in so often that we got used to it. After a while we didn't even try to spoil his fun. Sometimes we would leave door ajar so that Rocky, if he chose to do so, could achieve easy admission.

One day, I noticed a small amount of yellowing in the toilet water. Okay, somebody forgot to flush. Maybe it was me. I flushed.

A few days later I again noticed a tiny bit of yellow tint in the toilet water. A thought popped into my head but I quickly rejected it: *No, it can't be!*

But the evidence continued to accumulate that our crafty cat was up to a new trick. One day as I was walking past a bathroom door, I heard a tinkling sound. I stopped and turned just on time to catch sight of Rocky running out. When I went in to take a look, sure enough, there was a bit of yellowish fluid in the bowl.

This happened often enough that Joan and I knew what Rocky was doing, but he was clever enough to keep us from catching him in the act. Joan decided to keep a camera handy to record the scene in perpetuity, if she had the chance, and one night it happened.

I saw Rocky squeeze through the narrow opening of a bathroom door that was ajar and ran to inform Joan. We returned at just the right moment. I could hear a tinkling noise coming from inside as I pushed the door open.

Joan snapped the flash picture of a lifetime for a cat lover. I must be truthful and mention the fact that not everyone is impressed with the picture Joan took that night. Because of the use of the flash, it had dark areas around the edges. It showed too much plumbing to be considered artful. Technically, it had other imperfections.

But it had one important thing going for it: Unlike so many of the animal photos you see circulating on the Internet, this one wasn't posed or contrived. It was a candid classic.

The photo showed Rocky balanced with all four paws close together on the narrow toilet seat. He was facing sideways, looking in the direction of the camera. His rear end stuck out over the opening of the bowl, with his fluffy orange and white tail projecting upward at about a thirty-degree angle.

But the remarkable thing about the picture was not merely the evidence that Rocky had taught himself to pee properly. I had heard of other cats performing that feat, though, as far as I knew, the others were potty-trained. Rocky was self-taught.

The amazing and downright funny thing about that picture was the expression on Rocky's face. Just as I must have looked the first time Rocky broke in on me, Rocky appeared surprised…and embarrassed. The expression on his face seemed to say: "How could you humiliate me?!"

Afterward, when I flushed away the remains of what Rocky had done I noticed one more thing: Rocky's mannerliness. There was not a single stray droplet on the toilet seat.

Bathroom shenanigans was one of many things Rocky, the comedian, did during his seventeen years of entertaining my family. But there is one more thing I must mention about him: He was a pet of prayer.

Joan prayed many times through the years that God would send her an orange cat. To her, the arrival of Rocky was indeed an answer to prayer. It was a delayed answer because she had prayed for so long that she had almost given up hope. But even the delay turned out to be providential.

Joan was just entering a period of her life when she rarely felt well physically. Rocky, the clown, made her laugh – and the rest of the family too. He waltzed her through this trying time.

He showed up on a bus with children who had found him outside their school and wanted to adopt him. Their parents said no, so Rocky, a half-grown kitten at that time, went scrounging for food around the townhouse grounds.

Joan spotted him climbing the tree near our front door on his way to invading a bird's nest. She called me at work for my approval, but I think she would have gone after Rocky no matter what I had said.

The appearance of an orange kitten may not seem important as viewed from the perspective of a human life. But Rocky was extraordinary. I consider his timely arrival to be a piece of evidence of God's love for us.

He loved Joan enough to answer at exactly the right time a prayer that might have seemed inconsequential in comparison with major issues. Rocky helped Joan through that difficult time of her life. He turned out to be the most beloved pet we ever had.

So what? What's the big deal about the cat prayer? How do we know Rocky's appearance wasn't just coincidental?

Of course, we can't know for sure. But think about it: How many people do you know who have pet cats predominantly colored orange? Other than Garfield, I've never seen one. Well, actually, since Garfield was only a comic-strip cat, Rocky – who was funnier than Garfield, by the way – is the only one I've seen.

Rocky did encores for his humorous feats. When Joan would bring home the weekly groceries, Rocky would jump onto the kitchen table to watch her unpack them. When she was finished, she would place one of the empty bags on the floor next to the kitchen table with open end facing up.

Rocky would crouch on the edge of the table top and take a flying leap. He would disappear head first into the bag, but then the bag would topple and Rocky would flop over onto his back inside of it.

His favorite time was Christmas. He entertained himself all day by playing among the empty bags and boxes. If there was an empty box that had contained a large Christmas present, Rocky would nose around inside of it.

One year, an empty box was left standing on end about four feet in height. Rocky leaped onto the box, it tottered but didn't fall. Ta-DAH: Hilarious laughter and applause for the feline clown.

No cat I've ever seen could entertain the way Rocky could.

And how many of them have appeared out of nowhere at exactly the right time to be the pet someone needed and was praying for?

Rocky died in Joan's arms with a veterinarian attending. Afterward, in the car driving home, Joan said in a quietly reflective voice: "He really was a miracle cat. After the vet said he was gone, he was still purring."

Three years before Rocky's death at the venerable age of 17, a black and white kitten showed up at my workplace. I began feeding her in the evenings before driving home.

After a week or so, our daily meetings were anticipated. She would wait for me on a low tree limb near the parking lot outside my office. I would bring food and lay it between the roots of the tree. She would jump down and eat it. I began making it a habit to remember to make special trips to feed her on weekends and other days when it wasn't necessary for me to drive into the office to work.

I mentioned the kitten to Joan and we began to discuss adopting her. In addition to Rocky, we had three other cats. But Joan seemed intrigued by what I told her about the black kitten with white paws and fluffy white chest markings.

It was winter and cold weather was predicted. The night Joan and I decided I should bring the tuxedo kitten home, weather forecasts indicated a rare freeze in the Fort Lauderdale area. We feared the kitten would not survive the night.

When I took the food bowl out for her that evening, her response was different. Instead of jumping down and starting to eat, she walked over to me and meowed. She walked around me once, looked up at me and meowed again.

I picked her up and placed her on the unoccupied front seat of the car. But instead of resting there or fleeing in fear under a seat when the engine snarled, as most feral kittens would have done, she jumped onto the seat-back behind my head and wrapped herself around my neck. She stayed in that position for the entire drive. She was the softest, warmest scarf I have worn.

Joan was the primary care giver for our cats, so most of the rest of the elegant kitten's story came to me from her. Joan named her Victoria and referred to her as a princess. I called her Queen Victoria.

If Rocky was king of the house, Victoria was, indeed, the queen. Joan told me that of the dozen or so cats we had overlapping Rocky's lifetime, the only one he paid much attention to was Victoria.

"He trained her," Joan said.

Indeed, after Rocky's death in 2003, Victoria assumed his position of leadership among our felines. She is the one who orally reminded us of mealtimes and other regular feline festivities. The other cats hung around in anticipation that she would take the lead as Rocky always had.

The difference between Rocky and Victoria, as far as the other cats are concerned, is they don't like her as much as they liked him. But I do.

Since Joan's death more than a year ago, Victoria has been as important to me as Rocky was to Joan when she wasn't feeling well. I am healthy but there are times when I don't feel so hot. I am now 82 years old, and I miss Joan very much.

Victoria is 21, the oldest cat my household ever has had. Other than sleeping more than when she was younger, she doesn't show her age. She hollers at meal times and in between. She jumps onto the kitchen counter to make sure her demands are not ignored. Even at her venerable age, she is able to jump higher and farther than any of the other cats.

She lies next to me on a couch for hours while I sit to read the Bible, pray or watch TV. Sometimes she lies on my lap. Other times, she walks around on it before deciding to return to the couch beside me.

She's there for me during my time of need, just like Rocky always was for Joan. Recently, circumstances overwhelmed me to the point where I was distressed. I got out of my chair and knelt in front of it. While kneeling, I saw Victoria jump to the floor.

I assumed she had left the room rather than hang around a grieving man.

After I finished crying, I pushed myself up from the uncomfortable position on the floor. While rising, I felt my left arm bump against something. I looked down and got the surprise of my life.

Actually, two surprises.

First, the expression on Victoria's face showed sincere concern about me. The fact that she had sat by my side for a long time left no doubt about it.

A cat concerned about a distressed human being? Dogs certainly do not have a monopoly on sympathetic behavior. I vowed to stay by Victoria's side when the end of life came for her, as I thought it would, very soon.

I believe the appearances of both cats was divinely appointed – Rocky climbing the tree in front of our townhouse in a 36-unit complex and Victoria showing up on a tree branch outside my office when Rocky was aging.

Even the unusual tree limb appearances of the two hungry kittens don't seem coincidental to me. They were the only two cats out of nearly two-dozen to have been rescued by Joan and me from tree limbs.

CHAPTER SEVEN

'HERE AM I. SEND ME.'

I was 65 years old when I finally received word that I would see a Christian book of mine in print. By that time there had been a handful of basketball books with several more sports-theme books to come.

But my first love was reserved for Jesus Christ and my devoted Christian wife, Joan. I wanted to serve the Lord with my God-given writing ability, especially in the genre of fiction. So my excitement reached Super Bowl level when a respected Christian publisher offered me a traditional contract for my novel, at that time entitled *In His Steps Again*.

With a traditional contract, the book goes to stores, not just websites, and the author pays nothing. But in this particular case, the book was not to be published.

At a time when I was expecting word of a definite publishing date, I received a phone call from an editor at the publishing house. He told me that my book was one of several being cancelled because of a bad report about the previous group of fiction books the publisher had printed. Those books had lost money.

The next time something like this happened to me, the details were so ridiculous I had to laugh… when I wasn't shedding tears. By this time, I was 78 years of age and had 14 books in print. But none were as dear to my heart as *The High Sign*.

In 2005, two years after the *In His Steps Again* disappointment, I started reading the biblical book of Isaiah. I had been interested in biblical prophecy for a half-century, especially its prophecies pertaining to what the Bible calls the *end time* and the *last days* referring to the final era of world history. But I still didn't have clear understanding of it.

On the third day of this particular Bible study I came to Isaiah 6:8 and stopped abruptly. I re-read the verse in the New International Version:

Then I heard the voice of the Lord saying, "Whom shall I send? And who will go for us?"

And I said, "Here am I. Send me."

I had read these words many times, but this time they seemed to be addressed personally to me. After a few moments of prayerful thought, speaking as to the Lord himself, I said: "Here am I. Send me."

The next day, when I opened my NIV Bible, it seemed to be a different book. As I read chapter after chapter of Isaiah's prophecies referring to the *day of the Lord*, with second reference *that day*, I came to understand I was reading about the period of time preceding and including Jesus' prophesied Second Coming.

Had it been the prayer? Was God actually allowing me – a layman without a certificate of ordination in any Christian faith – to perceive meaningful patterns and details in the book that claimed to be his own?

This exciting new perspective on the prophetic parts of the Bible that had intrigued me since I was a teen-ager continued through that reading of Isaiah, the other Old Testament prophets, Jesus' own prophecy on the same subject in three New Testament books, and, finally, the end-time prophecies of Peter and the Revelation of John.

New bits and pieces of understanding continued to enter my mind as I read these same passages again, and have continued to expand my sphere of thought on this important subject to the present day.

My first book on the topic, *The Sign of his Coming*, was co-published by myself and Creation House in 2007.

Next, in 2013, came *What the Bible Has to Say about the U.S.A.* It was published by Lighthouse Christian Publishing.

After that, in January of 2016, came *The High Sign.* It summarized the first two books and added more evidential theme details than was to be found in those two combined. It showed linkage between 24 catastrophic events of the Bible, including the Second Coming.

Tate Publishing printed *The High Sign*, but then everything seemed to hit a dead-end.

At that time, I was in the recovery phase from two life-threatening illnesses and a reaction to the antibiotic Cipro that caused me more distress than the two combined. I'll discuss these in the next chapter. Here I want to emphasize details of the weird sequence of publishing mishaps I experienced.

I was recovering from my sicknesses when Joan contracted the deadly disease MDS. So after my own recovery, I did what I could do help her.

Tate Publishing went out of business one year after publishing *The High Sign*. Because of my health problems and Joan's, I had done no promotion work to stimulate book sales and so there were practically none.

But this depressing series of events took a humorous turn. One day early in the year 2017 I checked out the status of *The High Sign* on Amazon.com.

The book was listed as unavailable. No surprise, there, since it no longer was being published. But on that same page were listings for two unsigned, used copies of the book. Price being asked for the one was $630. The other, actually the first listed of the two, was tagged at $1,730.00.

I burst into laughter. This series of events had become so ridiculous that if I had included it in a novel, it would have drawn ridicule as an absurd line of plot.

But after the laughter came sorrow and disappointment. I knew Joan was near the end of life. Approaching the age of 80 myself, I had a hard time believing that God had not allowed Satan to have his way with my publishing mishaps.

What I did not realize was that God had plans for good that could not have been accomplished through the initial printing of *The High Sign* in 2016 by a financially troubled publisher. He had bigger plans for it.

During 2018, I had *The High Sign* published by Amazon.com in paperback and Kindle e-book. On the day this was written you could have purchased about 200 paperback copies of the book from Amazon for the same price being asked for a single rare copy of the book from its original edition published by Tate.

I think *The High Sign* is an important book, so I had it reviewed by the Online Book Club. The reviewer referred to me as a "Bible scholar" and gave the book the highest rating awarded by OBC. I was told by a representative of the book club that it is rare for any book to receive the maximum rating.

Since I began writing this chapter, *The High Sign* has been recognized as Book of the Day on the OBC's website. There have been opportunities for people to obtain the Kindle e-book free of charge for five days.

None of these things could have happened if Tate Publishing had not gone out of business. Tate's disappearance caused *The High Sign's* publishing rights to revert to me.

The breaking of the contract with me for publishing of *In his Steps Again* also turned out to be providential. I hold the rights to that manuscript now, too, and intend to publish it minus the final 70 pages. Those pages are anticlimactic. It will be a much better book without them.

But the legacy of *The High Sign* continues on.

A short time after it was Book of the Day for the Online Book Club, I received a communication out of the blue from URLink Publishing. They wanted to publish the book in a similar format to Tate but with a superior promotion plan.

Through all this, I have done little other than praying: "Here am I. Send me."

Since my declaration of commitment, both the writing of things that I did not understand for my first 65 years of life, and the three-fold avenues of publication for *The High Sign* – all of which previously were unfamiliar to me – must have been guided, in my opinion, by the Lord.

What other explanation could there be?

Could all of these things have been coincidental?

It is almost redundant to say I no longer feel disappointed about problems hindering the production of my Christian books. Instead, I am awed. I now regard God as infinitely wise and understanding of things that are beyond my powers of comprehension.

The series of events involving the cancelled publishing of two of my books, and the production of separate editions of one of them by three publishers within three years, cannot reasonably be assigned to the realm of coincidence.

This must be true because I hadn't had anything published previously by any of the three and had not sent a proposal to any of them. Somehow, all three decided to publish *The High Sign* even though they knew little or nothing about it.

It is no coincidence either, in my opinion, that these attention-attracting things are happening to what I believe is my most significant book.

CHAPTER EIGHT

SICKNESS THAT CURED ME

If I had to select one chapter of this book that is most climactic, this would be the one. What happened was more than surprising. To me it was astonishing.

The caveat is that this isn't fiction. I call it improbable non-fiction because it's the kind of thing that doesn't happen as often as once in a lifetime. To most people nothing like it ever happens at all.

It has happened only once to me. Within the plot of a novel, it would draw ridicule because it involves an unlikely sequence of events.

At the time when *The High Sign* first appeared in print, I began to have serious health problems. I'll describe the circumstances in narrative format with the understanding that seemingly unconnected pieces will be fitted together at chapter's end in what to me was a life-saving finish.

After my retirement from full-time professional journalism in 2007, Joan and I moved to a small town in Central Florida that hadn't changed much from the quaint place we recalled from our honeymoon 43 years earlier.

There was only one physician's office at that time in the little town. We joined Dr. H's patient list with the intention of remaining with him because he had a good reputation.

We probably would have stayed except for two things:

- There always seemed to be coughing patients in his tiny waiting room.
- We never got to see him, having to settle for dubious associates.

By dubious, I don't mean that they were lacking in medical education. To me, some of the things they said and did just did not make sense. And so I discussed the subject with Joan, who as a registered nurse was more qualified to make a decision than I. She said things weren't all that great in Dr. H's office, so we shopped around.

We settled on Dr. S, who was highly recommended. Perhaps because Joan tested out in better general health than I did, he saw me more often than he saw her. He usually assigned her to a female nurse-practitioner.

Maybe he thought women would be more comfortable discussing personal health issues with other women than with men. But Joan felt discriminated against. She raised the subject several times during the next few years, and I sensed that the time was coming when we would make another change.

Dr. S. kept careful records about my medical history on his computer. I also had a regular urologist, so Dr. S. did not keep track of urological issues. But on one of my visits he mentioned that he knew my urologist and he was getting ready for retirement. The implication seemed to be that maybe he wasn't as competent as he used to be. I filed that away in a mental folder.

I should have taken immediate action to change urologists.

I had been seeing Dr. S. for several years when one day, I remember it was a Friday, I began experiencing severe pain in the torso area. Preferring a regular doctor's office visit to a hospital emergency room, which would have been necessary on a Saturday when the doctor's office would be closed, I called to set up an appointment that afternoon with Dr. S.

He wasn't in the office that day, so I was slotted into the nurse practitioner's schedule.

The nurse practitioner didn't impress me any more than she impressed Joan. After examining me, she admitted she didn't know what was causing the pain but suspected an infection. She gave me a prescription for the antibiotic Cipro, and I began taking it that evening.

The next day the pain was much worse and I had lost my appetite. I was sure that if I forced myself to eat, I would regurgitate.

My urologist was affiliated with a hospital in Leesburg, but Joan had done volunteer work for more than a year at a hospital in Tavares closer to our home. She liked that hospital, so she drove me there and I spent the evening in the emergency room.

That weekend there was some testing, but apparently most of the skilled physicians were not on call. I spent the rest of Saturday and all day Sunday in a hospital bed, and did not eat a thing.

But I continued feeling worse and worse. It got so bad late on Sunday that I made a mental mistake that may have been attributable to rapidly deteriorating health. I forgot that my right arm was attached to a pole on wheels, from which dangled an IV that was dripping something into my right arm.

I arose to go to the bathroom and pulled the IV out of my arm. Within seconds, blood was spurting everywhere. I slipped in the blood and fell to the floor where a short time later I was found by an alert nurse. Her rapid application of first-aid to the torn arm might have averted emergency treatment.

By Monday morning, when I finally got to see a urologist, I had lost 10 pounds and was feeling terrible. The first thing the urologist, Dr. C., did was to change from Cipro to a milder antibiotic. I felt better immediately and from then on was able to eat and digest food.

My son Mike, an emergency medical technician, shook his head when he found out I had been given Cipro. "That's a dangerous antibiotic," he said. "It should only be given to patients in very critical emergencies."

Dr. C. ran some tests and learned that I had a kidney stone that was not giving me nearly as much trouble as had the Cipro. The next day, Tuesday, he removed the kidney stone. On Wednesday, I went home.

For weeks after this experience with what I suspected to be medical malpractice, I was angry. *None of it was necessary,* I told myself. *Just a great big expensive and dangerous hassle.*

But then something happened that changed my mind. To me, the problems caused by the Cipro, the bloodletting and the misdiagnosed kidney stone suddenly seemed minor by comparison.

During a follow-up visit with my new urologist, Dr. C., he discovered that my PSA tested very high. An x-ray revealed prostate cancer.

After showing me pictures of the half of the prostrate that was cancerous, he explained that there were new procedures for removing cancer from the prostate without chemo or radiation. Chemo and radiation were methods of treatment I had negative opinions about because of literature I had read by competent sources.

I was fortunate that Dr. C. agreed with me. If he had insisted on chemo or radiation, it would have been difficult for me to have said no.

Dr. C. said one alternative method involved burning the affected half of the gland. The other, he said, involved freezing. He said both were effective nearly every time but burning was slightly better than freezing.

When I learned that burning was thousands of dollars more expensive than freezing, I opted for the latter. He performed the procedure in an outpatient clinic and I drove myself home the same day.

I have been healthy for more than four years.

My opinion now is different from the first impression. At first, I considered the entire scenario to be a comedy of errors. All I had was a kidney stone, and I had had two of those before.

If Dr. S. had been in the office on the day I went in with middle torso pain, he probably would have diagnosed the kidney stone, because he had a record of my previous two stones in his computer file.

But if Dr. S. had taken charge, it would not have been necessary for me to go to the emergency room of what turned out to be the correct hospital to acquaint me with Dr. C, who was on call that weekend.

Dr. C. saved my life, not from the kidney stone, but from the cancer. I asked him about it, and he told me if it had gone undetected another year, I probably would have died.

As it was, the operation for removal of the cancer turned out to be the easiest procedure of all.

Were all of these things coincidental? I don't think so. I believe God arranged the negative-seeming circumstances to spare me from deadly consequences of a disease which at that time I didn't even know I had.

I have found out that the urologist I left to seek regular care from Dr. C was not even using the PSA test. He probably would not have detected the prostate cancer within the critical time period mentioned by Dr. C.

So I have been alive for at least three years, as far as I am concerned, on time purchased for me by God.

CHAPTER NINE

'I LOVE YOU'

Old photo albums have reminded me of the classic beauty of the woman who was my wife for 54 years. I've spoken about this and messaged about it, because I'm so amazed to have been her husband until her death three years ago.

But most of that memory is not about how she looked but how she loved. Several people who knew Joan have told me they sensed in her a great inner beauty. This is about the depth of that beauty at the end of a triumphant life. Those who saw her in her final week will know what I'm talking about.

After being spiritually reborn in her early thirties, Joan found the formula for basking in God's love: She prayed the Bible.

By that I mean she read it, meditated and talked to God about it, then read some more, meditated and discussed it with him some more. I have seen her spend hours doing this, and she did it more and more near the end.

She was intimately conversant with God. He spoke to her through the pages of the Bible, and she thought about it and talked it over with him.

The love of God in which she was immersed poured out of her onto others. As she aged, our cats spent more and more time in her favorite room whenever she was there.

Phantom, the youngest, so-named by Joan because, like a phantasm, this shy female black cat could do a vanishing act within the walls of our home, began to spend an hour a day with Joan. If I walked past the room during that hour, Phantom might become startled and flee.

It wasn't easy for me to explain that all I had done to frighten the cat was walk past. Joan would respond simply, "I love her."

Peering out her window at a flock of finches gathering at the bird feeder near the window, Joan would say, "I love them."

Joan's mood must have been contagious because I began telling her every time it entered my mind, at least once a day: "I love you."

Little did I know at the time that God had a plan for the great love he had implanted in Joan's heart. It would be the catalyst for a remarkable family reunion during her final weekend of life.

The pain and suffering escalated, but she loved more and more.

Joan was admitted to hospice on Monday, March 19, 2018. The doors are always open to visitors of hospice patients, so I spent many hours with her during her final week of life. Knowing she was near the end, I thought the chances of full family reconciliation were beyond hope.

On Friday, March 23, a hospice nurse, Judy Holtz, noticed the book I was carrying sheathed in leather. "I see you have a Bible," she said

"But it isn't even black," I said with humorous intent, holding out the book encased in cordovan.

"I know Bibles," Holtz said. "I hold a degree in Bible."

When Holtz found out about the one unfinished thing in Joan's life – reconciliation with our son Mike and daughter-in-law Angi – she left the room. She returned a few minutes later to inform us she had spoken on the telephone to Mike and Angi: "I told them where you are."

No longer able to eat, Joan was spending her last outing on the beautiful hillside porch abutting her room on Saturday when Mike and Angi walked in. Mike, whom I hardly recognized with his rustic beard, hugged me and then hurried out to hug Joan. There was an exchange of loving remarks that brought tears to my eyes, but the miracle was yet to be finalized.

I could not remember a time during Mike and Angi's 13 years of marriage when there had been complete amiability between Angi and Joan. There are a lot of jokes circulating about the misadventures of in-laws – especially mothers-in-law – but this situation wasn't funny.

It was about to end.

After hugging Mike, Joan noticed that Angi had taken a seat at a far corner of the porch. Though her voice was weak, it was clearly audible when she said to Angi: "I love you."

I arose from my chair next to Joan and offered it to Angi. Without a word, she walked to the chair. When she was seated on one side of Joan with Mike on the other, I stepped aside and said a silent prayer of thanks to God.

Before Mike and Angi left, I apologized to Angi for my role in the rift that had divided the family.

Joan fell into final unconsciousness late the next day.

The morning after, I knew, she would not awaken at her usual time of 4 a.m. After holding her hand through most of the night, I felt hers go limp in mine. She was still breathing, but I knew then that she would not regain consciousness in this world. It was, in fact, exactly 4 a.m. when I kissed her and whispered close to her ear: "I love you forever!"

I deliberately worded the sentence in the present tense.

I don't know if Joan heard me, but even though she was unconscious I do know that if God wanted her to hear, she did.

She died less than twelve hours later

According to the Bible, heavenly love is different from that of earth, but this I know: I shall have the opportunity to fulfill my final pledge to Joan. I shall love her forever in heaven with the purest kind of spiritual love. Since I am now convinced beyond doubt that the Bible is true when it states that the essence of God is love, I know this is an intention that pleases him.

On the way walking out of hospice I was given an armload of things that had been Joan's possessions. Without realizing it, I dropped a small bag. A nurse picked it up and handed it to me: It contained Joan's wedding ring.

"We managed to get it off her finger," the nurse said.

One look and I knew what to do: I removed the ring from the bag and tried it on the little finger of my left hand. It now rests within a half inch of my own wedding band, and I intend to keep it there.

After a graveside service the next Saturday, I had the opportunity to spend three hours alone with Angi. We talked and talked as if there never had been a problem between us, and there hasn't been one since that day.

During our conversation, I said to Angi: "Through the days before Joan died, she loved more and more."

Without hesitation Angi replied: "I noticed."

Since that time, I have exchanged visits with Mike and Angi several times and it is obvious that the three of us now love each other. It is what Joan had wanted most and what she herself accomplished during her final opportunity on this earth.

Her love multiplied itself among the rest of us.

The unlikeliness of this entire scenario was incomprehensible apart from a clear understanding of how far apart Joan and I had drifted from Mike and Angi.

I'm not even going to give serious consideration to the word *coincidence* concerning any of this because it was miraculous, no doubt about it in my mind. This was the work of God in response to prayers that had dwindled to near hopelessness.

God answered our prayers in his own time, which was, as it always is, exactly the right time.

The necessary presence of nurse Holtz wasn't coincidental. Neither was the dropping of the little bag containing Joan's wedding ring. God knew what I would do when I saw that ring, and arranged for it to be dropped right in front of me.

He wanted to see that ring – symbol of more than a half-century of a marriage that had grown increasingly strong through the final years – on my little finger. Now, more than three years later, it's still there.

And I have one other keepsake. While going through family memorabilia Joan had compiled, I found what I consider to be a postmortem message from her: *My husband Dave is best husband God could give to any woman.*

This message is now taped within my field of vision atop the computer monitor in which I am writing this: Joan, you were the best wife, and I can't wait to join hands with you as we rise together to our forever home on the Day of the Lord.

CHAPTER TEN

'HYMN'

As I write this, it has been only about a year and eight months since something happened that can be viewed as ordinary or extraordinary, depending on one's point of view.

At that time I began praying for something that was not high on my priority list. But I did pray this prayer numerous times. I prayed that my favorite vocalist, Sarah Brightman, would release a CD featuring Christian hymns.

I have CDs and DVDs of several Brightman concerts and studio sessions in which she sang Christian songs. One of these – her version of *Amazing Grace* – is so well done that I chose it for playing at both the graveside and memorial services for Joan. It received excellent response.

I know Brightman is a Christian because of what she said on PBS while discussing her song *Pie Jesu*. She asked the interviewer: "Would you like to know what the song's refrain means in English?"

The interviewer frowned and shook her head. Ignoring this cue, Brightman turned, smiled at the camera, and said: "Jesus is the Lamb of God who takes away the sins of the world."

I believed God was somehow involved in this, so I knew my request conceivably could receive a positive response. But I also knew that Brightman was going through a difficult time of life and had not done a concert or a studio CD of her usual stellar quality in nearly a decade.

The last thing she had recorded, to my knowledge, almost seemed to mock her extraordinary voice range. She consented to do a CD consisting of tuneless three- and four-note songs that I call jingalings.

So, assuming for the moment that God exists, here I was asking him for something that from his viewpoint probably did not even measure up to Joan's request for an orange cat.

To Sarah Brightman, if she decided to make a serious comeback, such a CD would have importance. But I probably had no business asking God for something that, even if I had a chance to meet Brightman personally and suggest it to her, might be considered effrontery on my part.

I did want to hear Brightman sing more Christian-theme material. But, even to me, this prayer seemed to have about as much hope of fulfillment as the sudden appearance of a septillion stars amid a universal vacuum (see Chapter 12).

If God was listening to this prayer, which even to me seemed unimportant, his response time probably would be at least as long as the time it took to approve

Joan's orange-cat prayer. I'd have departed from this earth by then.

Even so, I prayed persistently.

A few months later, I decided to do something I hadn't done in more than a year. I checked out Brightman's niche on Amazon.com.

And there, at the top of the page, I saw the word *Hymn*.

It is difficult to describe my response. I know that Joan was surprised but not shocked by the sudden appearance of an orange cat in our front yard. But I was more than surprised when I saw the word *hymn*. I was flabbergasted.

It was that very word I had been using in my prayer. And there it was, *Hymn*: The title of Brightman's new CD.

Basic dictionary definition for the word *hymn* is "a song of praise to God." The CD was appropriately named because hymns were what Brightman recorded for it.

If this story ended right there, I would have been happy because, though surprised, I was convinced that God must have heard my prayer. And He did not consider it too trivial a request to merit a positive response.

But that wasn't all. I played the CD and it did not contain a single traditional song that could be found in a Christian hymnal. The songs of praise to God were being heard for the first time by my ears. But the music was extraordinary.

Since receiving the CD, I have listened to it more than 150 times. It is a collection of some of the most beautiful music in the world, and only about half of it is in English – Brightman's usual 50-50 formula.

I have purchased and stored a second copy of the CD to play when my first one wears out. I have bought about a dozen others to give to Christian friends and family members. It has had the most impact on me of any musical recording.

Am I exaggerating? Not after having played *Hymn* many more times in twelve months than any one of my 200-plus other CDs in twelve (or more) years.

I'm not alone. A woman wrote in her review of *Hymn* for Amazon.com that she played the entire CD eight times on the day after receiving it.

My next door neighbor established a habit of playing it every evening.

My favorite song on the Brightman disk is entitled simply, *You*. It is a love song to the founder of the Christian faith and speaks these words in his praise: "I owe all I am to you."

Not *all-I-have*, but *all-I-am*.

I don't know if any of this was coincidence, or if my prayer had anything to do with the production of this CD. The project probably was undertaken before I started praying. But the naming of it may have occurred after my prayers began.

There are questions in my mind but these things to me are absolutes:

- *Hymn* is my favorite CD of all-time.
- To my ears, it is the most beautiful package of musical recordings ever assembled on a disk.

- It encouraged me at a time of my life when I needed it.
- I had been praying for it, even though I did not know a single song that would be on it.

Prayer seems to be working for me. Whether you call it providential or coincidental, the things for which I have been asking have been happening with remarkable consistency. So I have one more.

I have begun to pray that in the next life I will have the opportunity to sing harmony for Sarah Brightman for one song in a concert attended by angelic beings – including Joan, of course.

As for my ability to sing harmony, one year ago I was an off-key baritone. After joining the choir at my home church, First Baptist-Leesburg (FL), I was terrified upon being informed I was to be seated with the bass section between two men with voices much deeper and more melodious than mine. One of the two had been singing in choirs for more than 60 years.

A short time later my voice suddenly deepened by a full octave. Middle C used to be near the bottom edge of my range; now it's at the very top. I can go down and get those notes at the bottom of the bass clef, even though at times I have difficulty figuring out which ones to sing. I can hit deeper notes than one of the two men on either side of me and can match the other one's deepest bass.

Thank you, God, for changing my voice just when I needed it

Driving in my car while listening to *Hymn*, I interject notes that I think are harmonious. I think maybe I could sing at least a few notes of harmony with Brightman if given opportunity to do so.

Maybe this time Joan will be paying attention. Yes, it probably would amount to celestial showing off for the woman I love, who missed most of my terrestrial antics for her sake.

But maybe, just this one time, God will omit the punitive measures that he regularly doles out for excessive manifestations of pride. Especially if the song of choice is *You*.

Please, God. Let me do it for Joan.

(I think that You might enjoy it too.)

CHAPTER ELEVEN

CELESTIAL DECLARATION

My adventures on skis have been described in this book. Although I was proud of an initial mountain-busting descent, that and my fall on the last day of the trip represent for me a weekend of ups and downs. This chapter begins with the same trip to Vermont, entirely from a positive perspective.

After the first day of the trip we spent the night at a friend's mountaintop cottage. I think I have mentioned the three of us grew up and were living at the time in densely-populated areas. The cottage where we spent that night was isolated, but its solitary peacefulness wasn't the most memorable thing to me.

Urban/suburban areas along the eastern seaboard of the United States have weather that is overcast most of the time. Even on clear nights the moon and stars are not plainly visible. This is attributable to motor vehicle and air conditioner exhausts and other air pollutants.

That night at the cottage we developed stiff necks from star gazing. I saw celestial objects I had never seen before, and the familiar ones appeared much larger and more brilliantly beautiful than I had ever seen them.

We talked about it, but I can't recall the conversation. What I do remember is an overwhelming impression of the majesty of God. There is no other way to describe it, and, to me, no other reasonable explanation for the beauty we beheld.

Unlike most Americans of my generation who grew up during a time of exciting space exploration, I hadn't taken much notice of what the astronauts had described seeing during their celestial voyages.

But that night I recognized at least in part what the astronauts spent so much time talking about: The astounding order and beauty of a universe for which there seemed no other explanation than a creatively transcendent mind and power beyond description.

I had enjoyed personal and academic studies of astronomy. But, with few exceptions, the so-called constellations and signs of the Zodiac did not impress me because they didn't look much like the objects they were supposed to represent.

I could perceive the belt and sword of Orion the Hunter, but in the northern sky the only clear objects to me were the two so-called dippers. Somebody in authority, whose identity is mercifully unknown to me, had taken upon himself to muddle things up by saying they are not really dippers, but bears.

Bears? Yes, and the Latin word for bear is *ursa*, so the dippers are really not dippers at all but *ursas*. Correction: *ursae*.

Not in my astronomy book, they aren't. In my book they are dippers. They are ladles with curved handles and vessels for holding fluids.

That night in Vermont, the identities of the big and little dippers and all the other constellations disappeared among thousands of brilliant stars, most of which I hadn't seen before. These merged into a celestial ocean of incandescence.

Constellations: Where bears?

Houses of the Zodiac: Locked and the key thrown into the ocean in the sky.

Nothing up there I could see, other than celestial awesomeness. Yes, I know the word *awesome* has been overused by my sons' and grandsons' generations to the point of cliché.

But in this case, the word was appropriate. The sky that night was awesome. Stars appeared close enough to reach out and touch, even though I knew they were billions of miles away.

From where I stood, I couldn't even see the section of sky containing the lone object up there that looks entirely like the constellation it is supposed to represent: The Southern Cross.

I have read books about the constellations. One of these, *The Real Meaning of the Zodiac* by my former pastor, Dr. D. James Kennedy, convinced me that whatever objects are depicted by the glorious array of stars, the celestial canopy is God's work. It depicts the good news of eternal salvation through Jesus Christ.

Nothing else.

The cross is the only object up there representing anything that can be of use to any human being interested in experiencing the joy of an eternal relationship with the God who created all of it.

As I reflected on these things, I recalled other celestial phenomena. Solar and lunar eclipses interest me, even though lacking in color. For most of my career my job required late hours, so I witnessed more lunar eclipses than the average person, but few sunrises of the spectacular kind Joan enjoyed so much.

I have tried to make up for missing sunrises by taking late-afternoon walks to pick up the mail. The mailbox is about a quarter-mile from the front door, so I get to see a lot of wonderful sunsets. In Florida, many of these transcend in beauty even the work of a master painter.

Florida sunsets are most impressive with the full view of sky to the horizon afforded by an ocean, gulf or even a lake with calm surface giving a perfect reflection of the color-streaked sky.

Once, I saw an aurora borealis. That, too, was beautiful, although the featured colors were greens and blues instead of pinks, purples and oranges.

In Florida, there is one more regular celestial phenomenon. Clearing skies late in the afternoon of days in which there has been rainfall are conducive to the appearance of rainbows.

Except for the aurora borealis, I have seen all of these often. But I do not tire of looking for them again and again. From youth to old age, they have represented to me one thing and one only, expressed by a few poetic lines from the Bible:

The heavens declare the glory of God;
the skies proclaim the work of his hands.
Day after day they pour forth speech;
night after night they display knowledge.
There is no speech or language
where their voice is not heard.
Their voice goes out into all the earth,
their words to the ends of the world.
(Psalm 19:1-4, niv)

CHAPTER TWELVE

THE BASIC WORD

This chapter could have been injected anywhere in this book because it concerns a subject that I have investigated from time to time ever since receiving a D- on a college philosophy paper. I thought my grade should have been better.

(I hope the following does not give the impression that it has been motivated by the same prideful attitude that I confessed to in an earlier chapter. The next section of it does not have as its purpose to boast about myself but to show my writing credentials.)

I had won a writing award open to entries from everyone attending the university. I had been elected editor-in-chief of the student newspaper. My father, an editor for *The New York Times*, had praised one of my editorials, stating that if he had the task of editing it, he wouldn't have changed a word.

But the ultimate reason I thought my paper, entitled *God is More Probable*, was deliberately downgraded by the professor had nothing to do with my credentials. It was the note he scribbled across the top of it. He wrote that my paper was illogical.

Here is where readers of this book come into consultation. You may judge for yourselves whether my paper was illogical by the rest of this chapter, which employs the same reasoning as the original paper. Some recent material is inserted here that was not in the original paper, but the rationale is the same.

When I wrote that paper I had just aced a difficult course in logic. I had already created the beginnings of a statistical system based entirely on logic that later was used officially by professional basketball leagues on three continents, including the NBA.

My weaknesses are plentiful, but writing an illogical paper was not something I was susceptible of doing, even before reaching my 21st birthday.

I believed my paper was logical. It seemed to me that my D- was symptomatic not of my own illogic, but my philosophy professor's. He apparently did not like to be told by one of his students that the irreligious principles he espoused in his classroom lectures made less sense than Christianity or Judaism.

The issue, then and now, was whether we should believe the theory that human beings evolved from creatures of lower forms or that God made humans and other creatures with the inherent ability to reproduce the same kinds of creatures as themselves.

God created people and they had human babies. He created cats and they had kittens, dogs and they had puppies. We see this happening every day, and no one can demonstrate by scientific experimentation that it hasn't been going on in just this way for thousands of years. Scientific experimentation is relevant only to the present. It cannot tell us a thing about the distant past.

The Bible states ten times in its first chapter that God made creatures with the ability to reproduce after their "kind," not to evolve from one kind to another. Present observation of creatures having offspring of the same species is evidence for the truth of this biblical principle.

A foundational assertion of evolutionists is that the process took billions of years. However, my research has turned up no evidence that any creatures alive today are evolving into creatures of different species, even assuming that they did so in the past – a dubious assumption.

So, we have a companion to the elementary question of what launched the evolutionary process in the first place. If it's not happening any longer, why did it stop? By omitting God from their postulation, evolutionists create problems for their own theory.

The altered lifestyles of animals and birds to keep up with changing natural habitats have nothing to do with evolution. This is adaptation, the same phenomenon that can be observed in human beings. Move the comparatively short distance from the Rocky Mountains to the Pacific Coast and your way of life will change. But your membership in the human race will not expire.

If we accept the premise that evolution ever got started at all, the question of why it has halted remains without credible answer, even as evolutionists strive – in vain, it seems – to find evidence supporting the concept of a universe that generated itself from nothing.

No valid testing has been done to affirm the theory of evolution, even though a test is cited by evolutionists from the year 1953. That test did not prove that one species of animal can evolve into an entirely new species. Minor changes occur, which within a human context has resulted in racial differences. But they are races of people, not monkeys.

Charles Darwin, the most prominent evolution theorist of modern times, developed doubts about his initial postulations. Late in life, Darwin called his theory "grievously hypothetical." Speaking of his idea that the eye evolved by natural selection, he later said it seemed to him "absurd in the highest possible degree."[1]

The 1953 test could not be used to show how anything can suddenly appear where nothing was before – one of the key points of Darwin's famous book, *On the Origin of Species*. The experimental creation of a few amino acids during that test did not prove anything pertaining to the original Darwinian tenets of evolutionary theory because the 1953 test was not performed in a vacuum and did not involve originating something out of nothing.

It can be shown by scientific experimentation that nothing comes from nothing. So where did the

1 Darwin's Doubts, www.windowview.org

primeval oozing "pond" postulated by evolutionists come from? If there was no pond, where and how did evolution get its start?

As for the Big Bang Theory – and I am not talking about a TV show – in the absence of a creative hand, from what did the explosive substance originate? How was it ignited? Who lit the fuse?

Based on the outworking of the Second Law of Thermodynamics, the rate of deterioration of Earth in 26 scientific studies during the past half-century has been found to be a constant that would not have allowed life to exist on this planet more than 10,000 years ago.[2]

This is not a problem for Jews and Christians, because the span of time covered between the first and last pages of the Bible is about 6,000 years.

It is, however, a problem for evolutionists. Even if this calculation could be stretched by as much as five-thousand years, it wouldn't take a scientist to perceive the impossibility of closing the gap between 15-thousand and 15-billion years. A recent estimate of earth's age by a consensus of evolutionists is 15-billion years.

It is interesting that when I went to school more than a half-century ago, I was taught that the earth at that time was about three-billion years old. So it seems that the earth must have aged by twelve-billion years within a single generation. This brings us to

---

[2] Hugh Ross, The Creator and the Cosmos, Navpress, 1993, Pp. 118-121

another intriguing subject – the dating methods used by evolutionists.

During my research, I have learned that at least two of the high-school textbook "prehistoric" men were frauds: Piltdown Man and Peking Man. The recent application of paleontological science to the theme of evolution has been riddled by imposture and misrepresentation.

One "prehistoric" man turned out to have died only a year before his skeletal remains were found. A police investigation identified those dry bones as belonging to a man who had been a murder victim.

Evolutionists contend that the birth of the earth did not coincide with that of the origins of life. The Bible contradicts this in its first chapter. But even if we give credence to the argument that the existence of life is within the range of millions of years instead of billions, it makes little difference. The gap between ten-thousand and ten-million is too vast to bridge, even for an evolutionist.

Within the past fifty years the scientific community has identified more than forty parameters that must be kept within restricted range in order for a planet such as earth to support life. "Much less than one chance in a million trillion exists that even one such planet would occur anywhere in the universe."[3]

Even higher odds discount the possibility of a spontaneous (godless) Big Bang beginning for the

---

[3] Hugh Ross, The Creator and the Cosmos, NavPress, 1993, P. 144.

universe. Not only is the timing billions of years off target, but consider the essence: From where did enough combustible substance (hydrogen) come to provide the source for generation and sustenance of all those stars?

Our next question is a simple one: How many stars are in our galaxy?

According to recent estimates, there are anywhere from a few billion to hundreds of billions of stars in the Milky Way. The most credible estimate seems to be about 100-billion.

Final question: How many galaxies in the universe? Estimates vary, but this time they range from about two-trillion to ten-trillion galaxies.

Now, let's do the arithmetic: Multiplying the number of stars per galaxy by the number of galaxies yields an estimate of about one septillion stars. This is the number one followed by 24 zeroes.

Combustion, if and when it happens in the form that might be described as a "big bang," does not do so in the absence of prior existing combustible substances. It does not do so without a hand to light the fuse, in this case a septillion fuses – one for each star.

The simple truth is that, in the absence of a creative godhead, it is impossible to account for the existence of enough combustibles to explain the scorching durability of earth's solitary sun for 15-billion years.

Now consider the multiplication of this impossibility by one septillion, the number of stars.

That number times the tons of hydrogen needed to sustain each star for 15-billion years (another septillion) equals one grundecillion, which is the number one followed by 48 zeroes.

We have just described the "declaration" of the heavens based on Psalm 19, which was quoted at the end of chapter eleven. This declaration is based on odds involving a number with 49 digits to the left of the decimal point.

These are the odds against the spontaneous ignition and sustenance of one-septillion huge hydrogen furnaces (stars) for 15-billion years. In the real world, one grundecillion is so close to infinity as to be indistinguishable from it. There is simply no way to account for the pre-existence of all that hydrogen.

We aren't talking any more about an oozing "pond" of mysterious origin, but combustibles approaching the volume and weight that would preclude even the possibility of an original existence that was unexplained.

The most recent evidence supporting the premise of this chapter came from the scientific investigation of DNA. One conclusion of the investigating scientists was that all humanity is descended from a single man and woman. The Bible calls them Adam and Eve.

Extending their investigation to the animal kingdom, the scientists found that 90% of the animal species on earth today are descended, each from one pair of animals. These findings, by themselves, erase even the possibility of living creatures evolving from lower to higher life forms.[4]

David S. Thaler, an expert in genetics and microbiology, who participated in the research project, said, "This conclusion is very surprising, and I fought against it as hard as I could."[5]

The only explanation that fits the facts is the existence of an all-powerful, all-knowing being who made all of the elements present in the universe in immeasurable quantities and has been able to sustain a universal status quo to this day because his essence is, as the Bible implies, limitless.

Christians and Jews have a credible way of accounting for all of this: "In the beginning God created the heavens and the earth." (Gen. 1:1)

The online Yahoo answer to the question of how the sun consumes hydrogen as fuel mentions the same basic word: "…So you get fusion of these other elements that already have been created to make even heavier elements…"

The basic word is *created.*

---

4 Jonathan Cahn, Hope of the World Sapphires, June 2019 edition, P. 3.

5 Ibid.

CHAPTER THIRTEEN

BY THE NUMBERS

A few of the events described in this volume could have been coincidental, but the broad concept of coincidence for the entire project is opposed by exceedingly long odds.

It is the opinion here that God was involved in bringing about all of the seemingly unlikely outcomes described in this book. The only way to refute this is to find alternative explanations for each of them that make more sense than what is here written. An open-minded reader need look no further than the first chapter to recognize a divine presence.

How else can it be explained that within 24 hours of the deadline, a man who was unaware of my financial problem handed me a check for the exact sum needed to keep my family from becoming homeless, as described in Chapter One? What an extraordinary "coincidence" that he said before handing me the check: "God told me to give this to you."

Chapter Two: I can understand how questions might arise about an internal "conversation" in a hospital bed. It could be explained if I was deliriously ill. But I was actually lucid. Also opposing the delirious explanation was the fact that the attending doctor could not explain why severe pain from the illness disappeared overnight and medical testing the next day gave me a clean bill of health. When I entered the hospital I had severe abdominal pain. Overnight, *without* any medical treatment, but *with* much prayer support, I became pain-free and healthy.

Chapter Three contained descriptions of two incidents that may have involved angelic protection. I believe they were angels sent on missions from God. Evidence includes the fact that the first "angel" looked and moved like a human, though its see-through appearance eliminated the possibility of human identity. The amazing avoidance of collision by a speeding vehicle out of control on I-95 also seemed to be more likely angelic than coincidental. However, I confess that this chapter could not by itself stand as proof of the book's primary thesis.

The chapter (four) concerning pride also cannot stand alone. But it must be understood that the few incidents related in this chapter, though exemplary, did not come close to running the gamut of my offenses. Every time I can remember my ego surfacing in an ugly way there have been repercussions leading to embarrassment, and there were many such incidences unmentioned in this chapter. To me, this is a proof chapter: God hates pride. He did not like my attitude and disciplined me consistently through means, usually embarrassment, that hammered away at my problematic ego (Prov. 8:13).

Chapter Five was about my unprecedented prayer for God to protect a president of the United States. I prayed for President Ford for two straight days before a sharpshooter took dead aim at him from a distance of about forty feet. Did it just coincidentally happen that her brand new gun had such a terrible flaw in the aiming apparatus that her shot missed Ford completely? A career gun salesman told me he has never seen a new gun with such flawed sights.

Chapter Six: Other than Joan's insistence that Rocky was still purring after having been pronounced dead, there was nothing supernatural about him or his successor in our household, Victoria. The two cats both showed up in trees from which they were rescued by Joan and me. Rocky, in answer to Joan's prayer for an orange cat, arrived at the right time to help her cope with personal problems. Victoria has done the same for me. There may not be enough evidence to convince a skeptic, but these two precious animals kept Joan and me aware of God's loving presence for more than 30 years.

Chapter Seven: *The High Sign* is my favorite of the books of mine in print. Its writing was done through inspiration after I prayed the prayer of Isaiah 6:8: "Here am I. Send me." Its printing seemed miraculous, since Tate Publishing contacted me with an offer to publish it with no money required from the author. They did not know the title or contents of the book when they made this offer, and were unfamiliar with other things I had written. Once priced at $1,730.00 as a rare book, it has become one of the most widely read of the Christian books of mine in print, even though Tate went out of business before I began promoting it.

Chapter Eight is about the series of illnesses that wound up saving my life. Hospitalization for treatment of a kidney stone led to the discovery, through a series of tests, that I had second-degree kidney failure and a potentially fatal reaction to the antibiotic Cipro. Through the process of getting these things under control, I was introduced to a new urologist, and he saved my life. He discovered that I had prostate cancer and treated it in an outpatient clinic from which I drove myself home the same day. Since that day, five years ago, I have been in excellent health. Coincidence? Or was this scenario worked out by the Master Planner?

How do you describe God's love? I can't, but I believe its evidence is found in the description of my wife Joan in Chapter Nine. The closer she came to the end of life, the more she loved and loved and loved in a way imitative of His first love for her. Her final declaration of love resulted in a family reconciliation that prior to that moment had appeared all but impossible.

Chapter Ten: I asked God to inspire my favorite vocalist, Sarah Brightman, to do a CD consisting of Christian hymns. I persisted in this prayer until a few months later, when the thought crossed my mind to check out Brightman's space on Amazon.com. And there it was, at the top of the page – the word *Hymn*, title of Brightman's new CD of music praising God. I am not alone in believing it consists of some of the most beautiful music ever recorded. I play it several times a week and privately add bass harmony to some of the songs. Maybe this did initially seem coincidental, but now I don't think so.

Chapter Eleven: From childhood to the latter time of life, I have admired natural beauty. I am particularly impressed by the night-time sky, especially from a mountain-top perspective. Beginning with the fraternal trip to Vermont that was punctuated by some frantic skiing, I have had opportunity to observe the brilliance not so much of constellations but individual stars. To me, each star adds its own bit of support to the biblical declaration that the heavens proclaim the glory of God: The sun, moon and stars are his handiwork. (Psalm 19:1)

Even if coincidence might be the favored explanation for everything described in chapters one through eleven, the coincidental explanation is ruled out by a "grundecillion" of evidence emerging from chapter twelve. Odds against the spontaneous generation of enough hydrogen for the simultaneous consumption of one-septillion stars are so steep as to equate to a practical impossibility.

Evidence also overwhelmingly opposes the idea in vogue at present with evolutionists that the earth is 15-billion years old. Recent scientific studies have shown it could not be more than 10,000 years old.

Late in life Charles Darwin renounced his own theory of evolution. The alternative to his original theory that the universe began with a spontaneous big bang and then life inexplicably started to evolve, is contradicted at every point by more credible evidence: There was a Creator, whom the Bible declares to have been the eternally existing God.

CHAPTER FOURTEEN

FOREVER LIFE INSURANCE

I was trained in Evangelism Explosion, one of the most effective Christian outreach programs. But EE doesn't work without a preliminary question-and-answer session. The answer to this question is helpful: *Do you believe God exists?*

If God's existence is acknowledged, more detailed questions about him may be introduced for discussion, such as: *What about Jesus? Who was he and why was he important?*

Finally, the most important question: *If you died tonight and stood at the gatepost of heaven, and the gatekeeper asked you, "Why should I let you in?" what would you tell him?*

An answer focusing on the responding person, whether he/she has done enough good to compensate for the bad, indicates that person is not what the Bible describes as a born-again believer. The answer of a redeemed person will always focus on what Jesus has done, not on his/her own deeds.

To unsaved persons, the following scriptures and the attached explanations should be helpful in attaining the salvation they are seeking:

God demonstrates his own love for us in this: While we were still sinners, Christ died for us (Rom. 5:8). This Bible verse informs us that God loves us so much that he rescues us from the wretched sinful status that separates us from him. He did this through the sacrificial death of Jesus.

God made him who had no sin to be sin for us, so that in him we might become the righteousness of God (2 Cor. 5:21). While the sinless Jesus was dying on the cross the earth was darkened for three hours to show that God the Father, who dwells in pure light, had separated himself from the presence of His Son. Jesus took all human sin upon himself so the Bible could say of him that he had become sin. He died in our stead, and because of him God views us as righteous, if we ask him to forgive the things we have said and done with selfish intent.

Through the Spirit of holiness [Jesus] was declared with power to be the Son of God by his resurrection from the dead (Romans 1:4). The Holy Spirit came in power to identify Jesus as the Son of God. He did so by resurrecting him from the dead and enabling him to be seen by hundreds of people during a period of forty days before his visible ascension into heaven. Powerful earthquakes shook the world at the exact times of both his death and resurrection.

If you confess with your mouth, "Jesus is Lord," and believe in your heart that God raised him from the dead, you will be saved (Romans 10:9). This is a good test for the prospective Christian: If we truly believe that Jesus died for our sins and rose from the grave to prepare the way for us to do the same, we should be willing to declare that he is Lord.

Anyone who expresses a desire to trust the Lord for salvation should be given a chance to participate in a prayer including confession of sins and acceptance of God's offer of forgiveness and cleansing through Jesus (1 John 1:9). The confession part of the prayer may be silent. It is a personal matter between the repentant one and the Lord who suffered and died for him/her.

Those who make professions of faith in agreement with the quoted Bible verses should ask Christian ministers for the opportunity of being baptized into the Lord's death and resurrection to new life in him (Rom. 6:4). The word *baptize* comes directly from a Greek word meaning full-body immersion.

Another thing important to the new Christian is reading the Gospel of John, which was written for assurance of eternal life (John 20:31).

The new believer finally should set a goal to live a holy life moving forward from the salvation experience: *Now I have been crucified with Christ and I no longer live, but Christ lives in me. The life I live in the body, I live by faith in the Son of God, who loved me and gave himself for me (Gal. 2:20).*

The life a believer lives after this body decays will be one of perfect joy forever in the presence not only of loved ones from this earth but of the loving God who made it possible.

I have just given a summation of the gospel message. People who believe these things should live in a way demonstrating their faith.

But sometimes words can be inadequate. What were the full implications of Jesus' final statement before the death he suffered for our atonement when he said, "It is finished?" (John 19:10)

These three words are profoundly meaningful. They mean more than just that the physical suffering Jesus endured on the cross is over. They mean that he has finished all of the purposes for which he suffered and died.

He died for our full salvation and sanctification. This salvation involved him becoming sin in order to obtain for us forgiveness of all our sins (2 Cor. 5:21). Not only was the physical act of death finished, but also its spiritual aspects.

Jesus' physical suffering was defined by the blood he shed and the pain he felt, but these things in no way compared with the spiritual agony he went through in order to eradicate all of those sins.

The spiritual suffering that came upon Jesus for three hours while he hung on the cross was indescribable by the words commonly used in human language. This was because he drew to himself, like a cosmic magnet, the penalty accruing to trillions upon trillions of sins committed by billions upon billions of people.

For us, because of him, the burden is easy and the yoke light. For him, because of us, it was as weighty as the world.

For the reader who is uncertain about the truth of the principles outlined in this chapter, the 17th century mathematician Blaise Pascal offered a strategy for making a reasonable decision that has come to be known as Pascal's Wager.

Speaking of the God of the Bible, Pascal said every human being lives within a context in which he/she wagers everything important on a single lifestyle decision that, like it or not, must be made.

All of us, said Pascal, bet our eternal existence on a decision to live as if God exists or as if he doesn't. If we wager against God, and it turns out that he does not exist, our correct conclusion cannot be rewarded or even acknowledged by a nonexistent deity. For the same reason, if we live as though God exists, even though he doesn't, we lose nothing because of the error.

If God does exist, a way of life patterned after the belief that he does not results in infinite loss via hellfire, whereas a faithful way of life consummates with infinite gain in heaven.

So which option should we choose?

Seems to be a no-brainer, doesn't it?

CPSIA information can be obtained
at www.ICGtesting.com
Printed in the USA
BVHW080219091021
618423BV00007B/185

9 781647 537180